The Mysterious Disappearance of Amelia Earhart

by Myka-Lynne Sokoloff

EDUCATORS PUBLISHING SERVICE
Cambridge and Toronto

Series Authors: Kay Kovalevs and Alison Dewsbury
Series Consultant: Melinda Rice
Commissioning Editors: Sarah Russell, Bonnie Lass, and Sethany Rancier Alongi
Senior Editorial Manager: Sheila Neylon
Text by Myka-Lynne Sokoloff
Edited by Anne McKenna, Bonnie Lass, and Stacey Nichols Kim
Designed by Kathryn de Reus and Maria Vallianos
Photographic research by Emma Juniper

Making Connections program developed by Educators Publishing Service, a division of School Specialty Publishing and by Harcourt Education, a division of Reed International Books Australia Pty Ltd

ISBN 0-8388-3344-6

2010 2009 2008 2007 2006
10 9 8 7 6 5 4 3 2 1

Printed in China

Acknowledgments
The authors and publisher would like to thank the following for permission to reproduce the copyright material in this book.

Photographs
Australian Picture Library, pp. 5–6, 10–13, 16, 18, 21–25, 29, 31; Coo-ee Picture Library, p. 1, 4, 30; Courtesy of TIGAR, p. 24; Getty Images, pp. 14, 28; Library of Congress, Prints and Photographs Division (LC-USZ62-45002), cover; Purdue University, pp. 7–9, 15; TIGAR photo by P. Thrasher. ©1991 The International Group for Historic Aircraft Recovery. Used by permission. All rights reserved, pp. 26-27(all); USCG Historian's Office, p. 19.

Every effort has been made to trace and acknowledge copyright. The authors and publisher would welcome any information from people who believe they own copyright to material in this book.

Illustrations
Lewis Chandler

Contents

Introduction

In 1991, Dr. Tommy Love, a diver for The International Group for Historic Aircraft Recovery, was sitting on the sand on Nikumaroro, a Pacific island. A crab scuttled by and knocked over a leaf, revealing the heel of an old shoe. Could the shoe fragment help solve the puzzle of the disappearance of Amelia Earhart, the well-known aviator?

The mystery of Earhart's disappearance was more than fifty years old. In 1937, Earhart, her navigator Fred Noonan, and their plane disappeared while flying over the Pacific Ocean. More than three thousand people had taken part in a search, including at least sixty-five planes, a United States Navy battleship, an aircraft carrier, and destroyers. Neither the plane nor its passengers have ever been found.

Amelia Earhart

Who Was Amelia Earhart?

Amelia Mary Earhart is perhaps the most famous female aviator in history. Her life began in Atchison, Kansas. She was born on July 24, 1897, the eldest of two daughters of Edwin and Amy Earhart.

Earhart and her sister, Muriel, were fascinated by nature. Worms, crickets, toads, horses, and chickens were all interesting to these country girls. Even as a young girl, Earhart was ahead of her times. She preferred bloomers—loose shorts that gathered at her knees—to the long dresses and ruffled aprons most girls wore.

Amelia (at left) with her father, Edwin, and sister, Muriel

Earhart took risks when she was little. After a visit to a fair, the seven-year-old built a roller coaster in her yard. Riders could coast down wooden boards, which Earhart leaned steeply against the tool shed roof.

In high school, Earhart was an excellent student who enjoyed playing basketball. She got good grades, but if she wasn't interested in something, she made no effort to succeed.

Amelia Earhart, at the age of seven

Earhart as a young woman

Earhart became interested in flying planes when she visited a friend at an airfield in 1918. From this moment on, flying planes became a lifelong fascination for Earhart, and she was determined to find a way to fly. Some time later, she casually mentioned to her family that she wanted nothing more than to fly. Surprisingly, her father's only reply was to ask when she wanted to start.

Neta Snook (left) and Amelia Earhart

In January 1921, Earhart began to take flying lessons with Neta Snook, a woman just a year older than herself. Snook noticed that Earhart often made mistakes when she flew.

Earhart took off one day for a fairly long trip without checking the fuel tank. She assumed that it was full, since the owner always filled it. In fact, the tank had not been filled. She landed safely, but she had taken a risk. Earhart also confessed to Snook that she sometimes daydreamed when flying.

In July 1921, just before her twenty-fourth birthday, Earhart bought a plane. The purchase worried her friends. The plane had a small, unreliable engine. However, Earhart liked it because she could pick up the tail and move the plane by herself. She named it "The Canary."

Within days, Earhart had her first plane crash. Earhart was fine, but the propeller and landing gear were smashed. Earhart didn't let that deter her.

In October 1922, Earhart went on a solo flight. Without telling anyone, she decided to see how high she could fly. Could she top the women's altitude record? She broke the record, reaching fourteen thousand feet.

Earhart, around 1920, with her first plane, which she called "The Canary"

Crossing the Atlantic

In 1927, U.S. aviator Charles Lindbergh successfully crossed the Atlantic Ocean on a nonstop flight from New York to Paris. A few months later, Earhart was invited to be the first woman to cross the Atlantic by air, as a passenger in an airplane called the *Friendship*.

The *Friendship* left Newfoundland, Canada, on June 17, 1928. Earhart was accompanied by pilot Bill Stultz and mechanic Slim Gordon. Pontoons—supports to help the plane float—had been added so that the plane could take off from and land on the water. After a bumpy start, the three passengers made a successful crossing to Wales, in Great Britain. Earhart became a celebrity. She went to lavish parties and even danced with the Prince of Wales.

The crew of the *Friendship*, after their arrival in Britain. Earhart is in the middle, with Gordon to her left and Stultz to her right.

Although Earhart was considered the commander of the *Friendship* flight, she never actually flew the plane. Some people criticized her for being just a passenger. Earhart felt that she had something to prove to the world. She would show that women could be pilots as well as passengers. In 1928, she became the first women to complete a solo round trip across the United States.

Earhart was becoming more and more famous, with the help of a man named George Palmer Putnam. Putnam set up interviews and arranged her publicity. Over time, Putnam and Earhart became very close, and in 1931, they got married.

Amelia Earhart and her husband, George Putnam

Earhart after landing in Ireland, following her solo flight across the Atlantic in 1932

In May 1932, Earhart attempted a solo flight across the Atlantic, again leaving from Newfoundland. The trip was not a smooth one. The altimeter, which measures a plane's elevation, or height, failed to work. Earhart flew for some distance in thick fog, unsure how high she was above the ocean. At one point, she could see flames shooting from part of the plane. Still, she landed safely in Ireland, becoming the second pilot and first woman to fly solo across the Atlantic.

After Earhart's transatlantic flight, she met with royalty all over Europe. Back home in the United States, she received the National Geographic Society Medal from President Hoover. Soon after, she became the first woman to receive the Distinguished Flying Cross from Congress. Numerous other awards increased the public's excitement about Earhart the aviator.

In her trademark brown bomber jacket and with her cropped hair, Earhart became a familiar sight in newsreels around the world.

In August 1932, Earhart set the women's United States transcontinental speed record, flying nonstop from Los Angeles, California, to Newark, New Jersey, in nineteen hours and five minutes.

Soon, Earhart would no longer be the first *female* pilot to achieve many important aviation goals. She would be the first *pilot*, male or female, to accomplish several feats.

In 1935, Earhart was the first flier to cross the Pacific Ocean from Hawaii to California. As she prepared for this trip, the United States Navy said that her radio equipment was not safe. But Earhart ignored this advice. In the same year, she was the first person to fly solo between Los Angeles, California, and Mexico City, Mexico. She was also named America's Outstanding Airwoman and was the first woman to compete in the National Air Race.

Earhart at the controls of her plane

Around the World

Earhart's plane is surrounded by hundreds of people after her 1935 flight from Hawaii to California.

Earhart's fame continued to grow. Tens of thousands of people mobbed her at public events. She began to plan a round-the-world trip. Earhart was not the first flier to attempt this feat. It had been achieved by a United States Army service team in 1924, and by a pilot named Wiley Post in 1931 and 1933. But Earhart wanted to fly close to the equator. This would make her journey the longest ever flown. And she would be the first woman to make the trip.

Earhart's plane was damaged on her first round-the-world attempt.

In March 1937, Earhart began her round-the-world trip, flying a custom-built, twin-engined Electra plane. She flew west from Oakland, California, to Honolulu, Hawaii, in record-setting time. On the second leg of her journey, she left Hawaii for Howland Island in the South Pacific Ocean. The plane was badly damaged on take-off, however, and although Earhart was fine, the trip was called off.

After repairs to the plane, Earhart began a second round-the-world attempt. This time she planned to fly from east to west. The official start was in Miami, Florida, on June 1, 1937. She changed radio equipment, and she decided to leave the Morse code equipment behind. Were these decisions wise? Time would tell.

Final Moments

On July 2, 1937, Earhart and her navigator, Fred Noonan, were coming to the end of their twenty-seven thousand-mile trip around the world. They had already flown twenty-two thousand miles in the Electra. They stopped often to greet cheering masses of people.

They expected to complete the 2,556-mile trip from Lae, in New Guinea, to Howland Island that day. They were prepared in case of mechanical problems. The plane could fly with one engine, if necessary. There was enough fuel to carry them four thousand miles. However, they never arrived at their destination.

This map shows the planned route of Earhart's second round-the-world attempt.

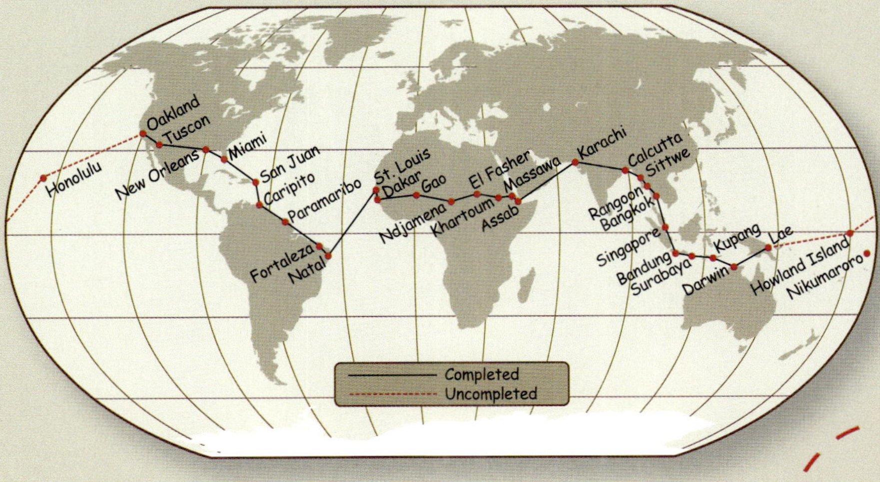

A United States Coast Guard ship, *Itasca*, was stationed near Howland Island, roughly one thousand miles southwest of Hawaii. It was there to help guide the fliers for this leg of the trip. The plane and the ship were in frequent radio contact. But at 7:42 a.m., *Itasca* received this message from Earhart:

"We must be on you but cannot see you, but gas is running low. Been unable to reach you by radio. Flying at a thousand feet. Only half-hour's gas left."

It was the last confirmed message received from Earhart.

Earhart and Noonan board the plane for their flight.

The U.S. Coast Guard ship *Itasca*

No one knows exactly what happened in the cockpit of Earhart's plane during its final moments in flight. Radio messages suggest a couple of possibilities. Perhaps Earhart thought she was running low on fuel and decided to land the plane to avoid a crash. Perhaps she wandered off course and could not find a place to land in the vast Pacific Ocean.

What Happened to Amelia Earhart?

"When I go, I'd like best to go in my plane," Earhart reportedly said on several occasions. It appears that her wish came true. Earhart, Noonan, and their plane have never been found. Many theories suggest what may have happened to them.

Did the Fliers Go Off Course?

Fred Noonan was on board specifically to chart the plane's course. Today, navigation is more accurate than it was in 1937. Noonan used charts, but he did not have the most current ones available in 1937. In fact, *Itasca* had corrected the charted position of Howland Island a year earlier. Earhart's charts did not show the change. The fliers could easily have gone off course.

Earhart's maps did not show the correct location of Howland Island, which is shown here.

Was Radio Trouble to Blame?

Experts agree that Earhart was not skilled at radio communication. Her plane had new radio equipment and Earhart was not familiar with it. She failed to broadcast her location twice an hour as planned. There was confusion about what radio signal to use. A careful look at film of Earhart leaving on her final leg shows that the plane may have lost its radio antenna during take-off.

Perhaps most important was Earhart's decision not to carry Morse code signaling keys. Morse code was a more reliable signaling technology than radio, but both Earhart and Noonan were reported to be slow at using Morse code and had decided to leave the equipment behind. This left them with only radio communication. If this failed, there was no other way to communicate with people on land.

Earhart and Noonan with a map of the Pacific that shows the planned route of the last flight.

Did the Plane Run Out of Fuel?

A simple explanation for Earhart's disappearance is that the Electra ran out of fuel and crashed into the Pacific Ocean. But Earhart carried more than enough fuel to reach Howland Island. So what made her report that she was nearly out of gas? Perhaps this message was misunderstood. Some people believe she said she was a half hour away, not that she was running low on fuel.

A group called Nauticos has carefully examined the messages between Earhart and *Itasca*. They used the radio details to narrow their search to the waters around Howland Island. They believe that they will find the craft by using underwater sonar for the search. So far, they haven't found anything.

Earhart's niece and nephew listened to radio reports of the search for their aunt on July 5, 1937.

Did the Fliers Survive an Emergency Landing?

There are reports of radio signals received after the Electra's disappearance. For a few days after the disappearance, people heard bits of conversation on the radio. A woman's voice was heard speaking English. These snippets suggested that Earhart and Noonan were alive on land somewhere. But no one could signal the fliers. Did they survive an emergency landing? No one really knows.

Liberty Scout Wins 2d Delaware

THE EVENING SUN

Sports 6 Star SATURDAY SPECIAL

BALTIMORE, SATURDAY, JULY 3, 1937

Budge Wins 3 Titles, Wimbledon Record

RADIO MEN HEAR EARHART'S VOICE FAINTLY ASKING FOR AID

ELLABEE TAKES FIRST DELAWARE RACE BY NECK

Trailing New Suspects In Kidnap Case

BUDGE-MAKO TEAM CAPTURES DOUBLES TITLE

FLASH

BRITAIN WARNS AGAINST SEIZURE OF SPANISH LAND

Denhardt Is Accused In Party Girl's Death

WEAK MESSAGE PLACES PLANE NEAR EQUATOR

2 GEORGIANS WIN 100-METER TEST

AB WRIGHT HITS HOMER, CLEARING BASES FOR BIRDS

Delaware Pk. Chart

BASEBALL

OUSTS 3 LEADERS IN STEEL WALKOUT

SHOE STAND YIELDS $450,000 IN BONDS

This newspaper, dated July 3, 1937, reported that radio signals from Earhart were heard after her disappearance.

Did Earhart Take on a New Identity?

Others think that Earhart returned to the United States and took on a new identity. They argue that this happened so that she would no longer need to live such a public life. Other related theories exist, but there isn't much evidence to back them up. Given how well-known Earhart was, it seems unlikely that she could hide in this way.

Was Earhart a Spy?

Earhart's flight took place not long before the beginning of World War II. Some people think that Earhart was spying on Japan for the United States. Perhaps she was captured and killed. Today, there is little evidence to suggest she was a spy.

MISS EARHART FORCED DOWN AT SEA, HOWLAND ISLE FEARS; COAST GUARD BEGINS SEARCH

ROUTE OF EARHART PLANE IN PACIFIC

The flier took off from Lae, New Guinea, at 8 P. M., New York time, Thursday, and was in the vicinity of Howland Island at 4:43 P. M. Friday, when she was in communication with the Coast Guard cutter Itasca. She had intended to fly from this island to Honolulu and thence to the mainland for the completion of her world tour.

FUEL HAD RUN LOW

Fliers Were Near Goal When Last Reported but Saw No Land

PLANE EQUIPPED TO FLOAT

Has Sealed Gasoline Tanks and a Rubber Lifeboat for Emergency at Sea

RADIO BELIEVED HEARD

Los Angeles Amateurs Pick Up Weak Signals on Frequency Assigned to the Plane

A newspaper report announces the search for Earhart's plane.

Searching for Clues

At different times in the past, people searching for the lost aviators heard stories of Western pilots seen on small Pacific islands. Some reports told of bones found on different islands. There is no solid evidence that the sightings were of Earhart or that the bones were hers.

In July 1960, an airplane generator was found in the South Pacific, which may have been from Earhart's plane.

Members of TIGHAR on Nikumaroro in 1991

One group that continues to look into the mystery is The International Group for Historic Aircraft Recovery, known as TIGHAR (say: *tiger*). Group members have visited the island of Nikumaroro several times. The island, which was named Gardner Island in the past, is called Niku for short. It is part of the Phoenix Islands of the Republic of Kiribati.

Nikumaroro lies about three hundred miles southeast of Howland Island. In searching the island in 1989, the group found some plane parts. Tests show that they did not come from a military plane. The parts had the same type of rivets used on the Electra. The plane parts were found about two miles from where Dr. Tommy Love would find the heel of a shoe in 1991.

TIGHAR sent the heel Dr. Love found to the company that made the shoe. After careful examination, the company found that the shoe piece had been made in the 1930s and it had come from a woman's shoe, probably a large-sized one. It was from a style Earhart was known to wear when she flew, and she also wore a large shoe size.

Few, if any, people lived in the area at the time of Earhart's disappearance. And even fewer of them wore shoes, let alone a style and size worn by the flier.

Right, the heel found by Dr. Love. Below, parts of the shoe were pieced together to show the original form.

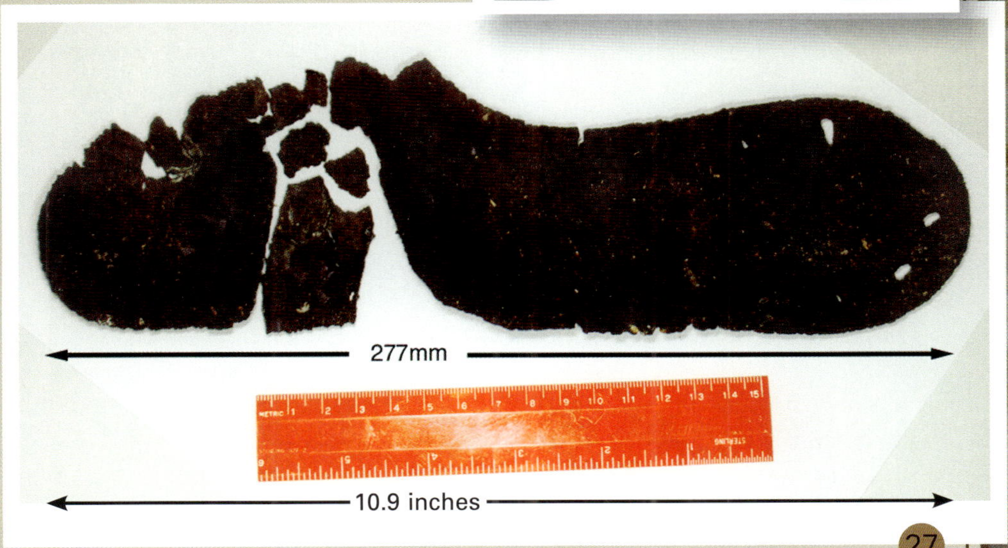

Conclusion

Hopes were high in 1937 when Earhart and Noonan left on their round-the-world trip. George Putnam, Earhart's husband, had been asked about their chance of success. "The chances? I don't know. I can only hope," was his reply. "I do know that if courage and fine determination can pull them through, both they have in abundance, and will return."

Even when Earhart had been missing for twenty-four hours, Putnam remained optimistic. He may have believed at one point that Earhart had been on a government mission and was captured by the Japanese. Even he eventually gave up hope of ever finding his wife.

Earhart in Hawaii in 1935

The fascination with Earhart's disappearance continues today. How much did pilot error add to the events that unfolded? What part was due to failed equipment—or lack of equipment?

Some of the theories about her disappearance seem unlikely. Will the mystery of Earhart's disappearance be solved one day? Some searchers believe they have found important evidence. Others are not sure. You have the facts now. What do you think?

The plane that Earhart flew from Hawaii to California in 1935 is now on display at the National Air and Space Museum in Washington, D.C.

Timeline of Events

July 24, 1897 Amelia Earhart is born in Atchison, Kansas, United States.

1921-1922 Earhart takes flying lessons with the pilot Neta Snook.

October 22, 1922 Earhart sets the women's altitude record of 14,000 feet.

June 1928 Earhart is the first woman passenger to cross the Atlantic Ocean by plane.

October 15, 1928 Earhart completes the first solo round-trip transcontinental flight across the United States by a female pilot.

July 5, 1930 Earhart sets the women's speed record of 181 miles per hour in a single-engine plane.

February 7, 1931 Earhart marries George Putnam.

May 20-21, 1932 Earhart flies solo across the Atlantic Ocean.

August 24-25, 1932 Earhart sets the women's nonstop speed record for crossing the United States.

January 1935 Earhart is the first person to fly from Hawaii to California.

April 1935 Earhart is the first person to fly from California to Mexico.

March, 1937 Earhart begins her first round-the-world attempt.

June 1, 1937 Earhart and Noonan begin their round-the-world flight.

July 2, 1937 *Itasca* receives the last known radio contact from Earhart and Noonan.

July 18, 1937 U.S. Navy stops searching for the missing aviators.

Index